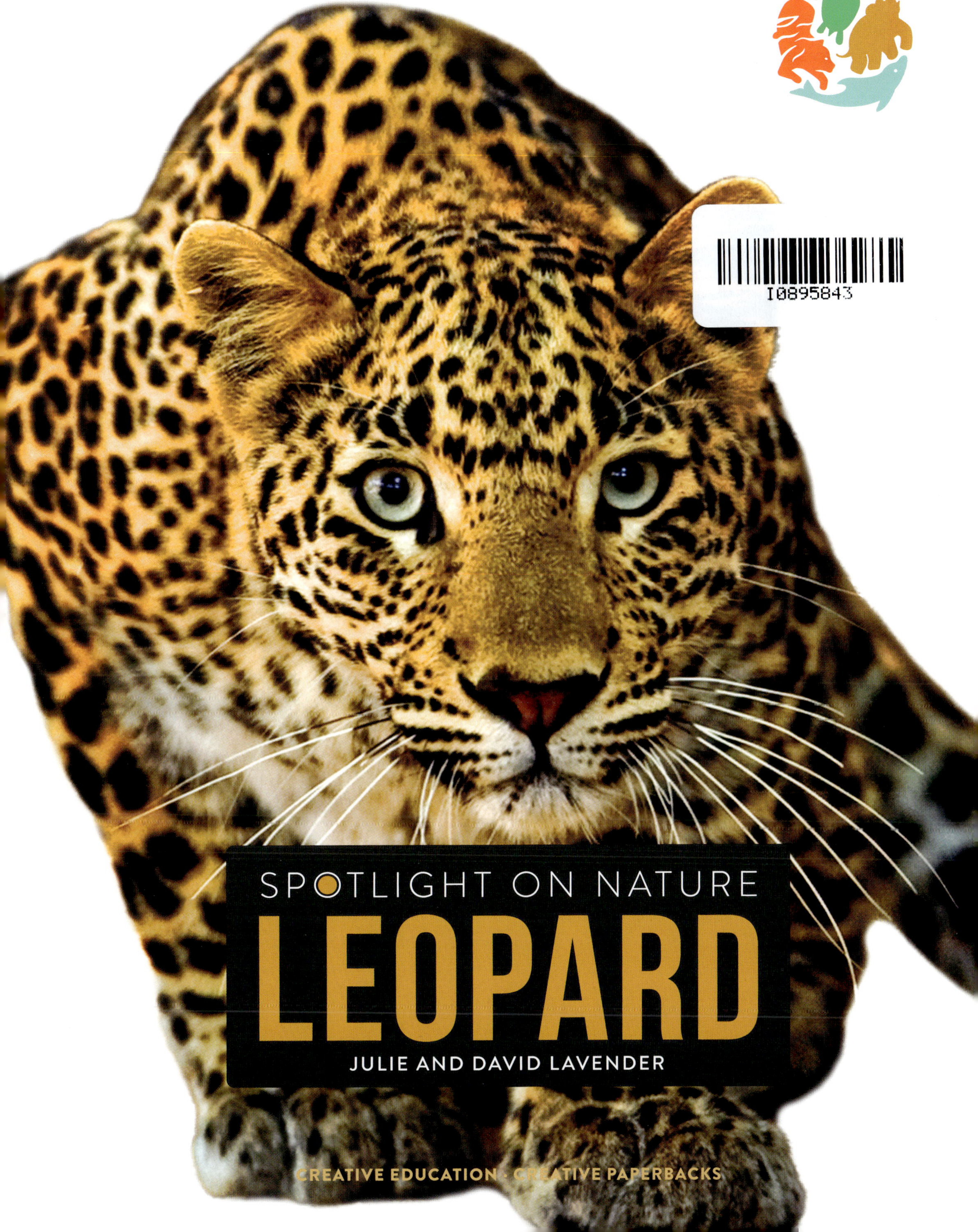

SPOTLIGHT ON NATURE

LEOPARD

JULIE AND DAVID LAVENDER

CREATIVE EDUCATION · CREATIVE PAPERBACKS

Published by Creative Education and Creative Paperbacks
P.O. Box 227, Mankato, Minnesota 56002
Creative Education and Creative Paperbacks are imprints
of The Creative Company
www.thecreativecompany.us

Design by Jennifer Bowers; production Beeline Media & Design, Inc.
Art direction by Tom Morgan
Printed in the United States of America

Photographs by Dreamstime (Dennis Schaefer, Grayfoxx1942, Andrea Poole), Getty Images (Aditya Singh, Leopardinatree, skynesher, Steve Adams), Shutterstock (EcoPrint, MeadwayK, M Rutherford, slowmotiongli, RachenStocker, Grindstone Media Group, Eric Isselee, Xyo, james_stone76, Stu Porter, Pavith Malaka, Albie Venter, MrNoe, rajvish stock, Ondrej Prosicky, ylq, kyslynskahal)

Library of Congress Cataloging-in-Publication Data
Names: Lavender, Julie, author. | Lavender, David R., 1961- author.
Title: Leopard / Julie and David Lavender.
Description: Mankato, Minnesota : Creative Education and Creative Paperbacks, [2025] | Series: Spotlight on nature | Includes bibliographical references and index. | Audience: Ages 10-13 | Audience: Grades 4-6 | Summary: "Prowl the leopard's world with our new book for mid-level readers. An engaging narrative of a leopard family, captivating images, infographics, milestones, and resources make it perfect for young nature enthusiasts"-- Provided by publisher.
Identifiers: LCCN 2023059467 (print) | LCCN 2023059468 (ebook) | ISBN 9798889891956 (library binding) | ISBN 9781682775684 (paperback) | ISBN 9798889892038 (ebook)
Subjects: LCSH: Leopard--Juvenile literature. | Leopard--Life cycles--Juvenile literature. | Leopard--Behavior--Juvenile literature. | Leopard--Amur River Valley (China and Russia)--Juvenile literature.
Classification: LCC QL737.C23 L3687 2025 (print) | LCC QL737.C23 (ebook)
| DDC 599.75/5415--dc23/eng/20240125
LC record available at https://lccn.loc.gov/2023059467
LC ebook record available at https://lccn.loc.gov/2023059468

CONTENTS

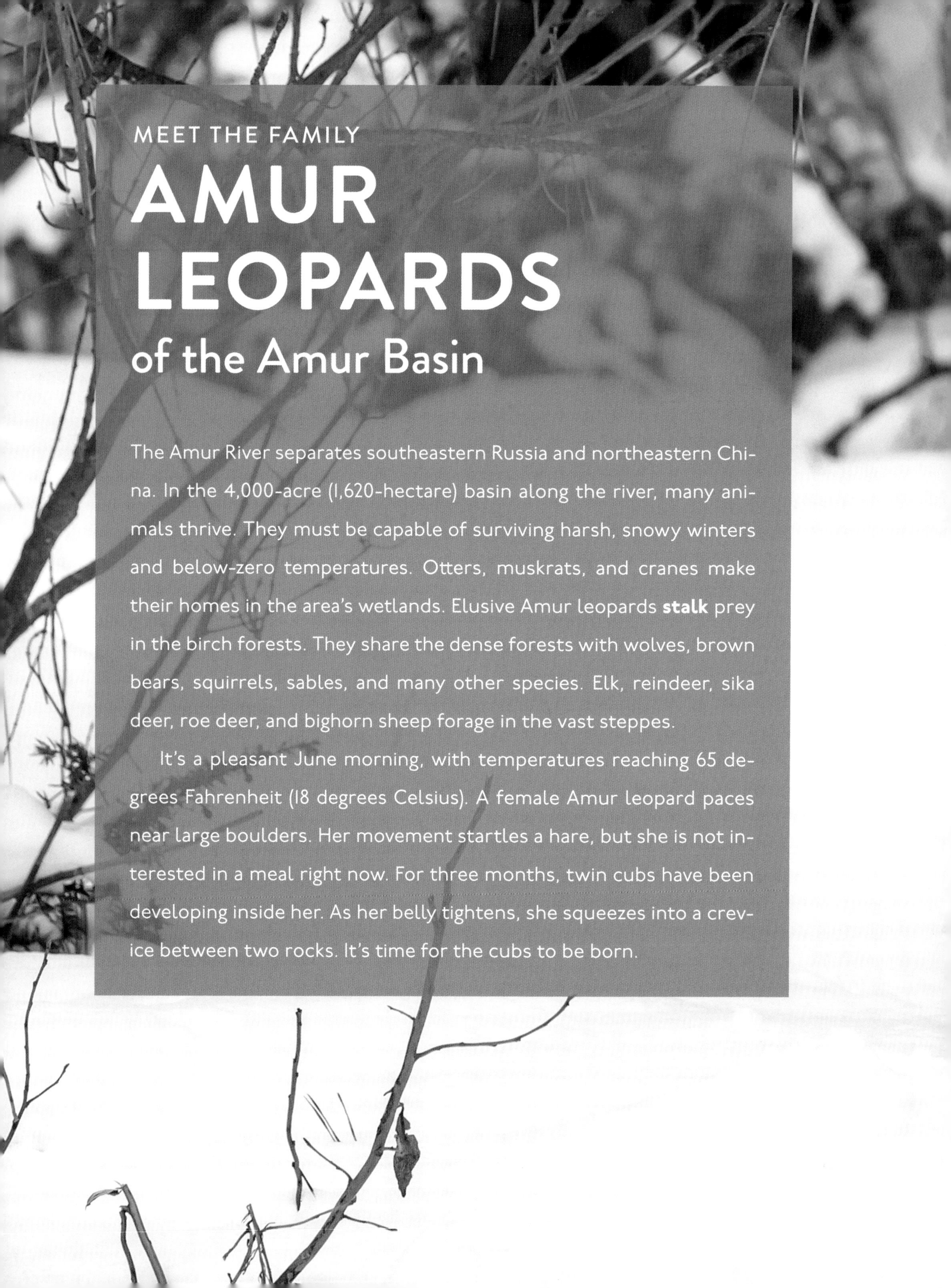

MEET THE FAMILY

AMUR LEOPARDS

of the Amur Basin

The Amur River separates southeastern Russia and northeastern China. In the 4,000-acre (1,620-hectare) basin along the river, many animals thrive. They must be capable of surviving harsh, snowy winters and below-zero temperatures. Otters, muskrats, and cranes make their homes in the area's wetlands. Elusive Amur leopards **stalk** prey in the birch forests. They share the dense forests with wolves, brown bears, squirrels, sables, and many other species. Elk, reindeer, sika deer, roe deer, and bighorn sheep forage in the vast steppes.

It's a pleasant June morning, with temperatures reaching 65 degrees Fahrenheit (18 degrees Celsius). A female Amur leopard paces near large boulders. Her movement startles a hare, but she is not interested in a meal right now. For three months, twin cubs have been developing inside her. As her belly tightens, she squeezes into a crevice between two rocks. It's time for the cubs to be born.

CLOSE-UP

Whiskers

Leopards have specialized whiskers, called mystacials, on their upper lip. These whiskers have sensory nerves that help the leopard know the position of the prey it captures. The whiskers help the leopard make a precise bite for a quick kill.

CHAPTER ONE

LIFE BEGINS

Leopards are one of the five mammals in a group called big cats. Also in the group are tigers, lions, and jaguars, which are larger than leopards. The fifth big cat, the snow leopard, is about the same size as a leopard.

The leopard lives in more places on Earth than any other big cat. Nine **subspecies** make their homes on the continents of Africa, Europe, and Asia. Leopards live in a range of habitats including grasslands, forests, mountains, deserts, and rainforests. They are **opportunistic** hunters that eat just about any animal they can find. Their favorite foods seem to be animals with hooves such as antelope, deer, or gazelles. Leopards also eat monkeys, mice, rats, reptiles, birds, fish, and even insects. Sometimes, the expert hunters become the hunted. Lions, tigers, hyenas, wolves, crocodiles, and pythons sometimes prey on leopards.

AMUR LEOPARD MILESTONES

DAY 1

- Born
- Thin fur with faint spots appears gray
- Blind and toothless
- Weight: 21 ounces (595 grams)

1 WEEK

- Lives on mother's milk
- Opens eyes
- Crawls
- Mews softly

Most leopards have light-colored fur in shades of yellow and orange. Black splotches shaped like roses cover the leopard's body. Scientists call these spots rosettes. Adult female leopards range in size from 45 to 132 pounds (20 to 60 kilograms). Adult males weigh 80 to 165 pounds (36 to 75 kg). Measured from the tip of the tail to the nose, leopards can be nearly 10 feet (three meters) long. That's longer than two fifth graders stacked head to toe.

Adult leopards are solitary animals. They come together only for courtship and mating. A mother leopard gives birth to one, two, or three cubs in a litter. Each cub weighs between

CLOSE-UP

Helpful tongues

Leopard tongues have specialized bumps. Each bump has a sharp hook. This feature helps leopards groom their fur. It also helps them get every last morsel of meat off the bones of their prey.

FEATURED FAMILY

Welcome to the World

The mother leopard reclines on the ground and stretches her legs uncomfortably. She rolls on her back and pants heavily. Soon, the first cub, a female, is born. The mother licks her clean. Two hours later, after more contractions, another cub is born. This one is a male. Once again, the mother licks the newborn clean. The cubs' fur is thinner than the mother's. Faint spots give them a grayish color. The twins are weak, wobbly, and toothless. They weigh just over one pound (0.5 kg) each.

one and two pounds (0.5 to 1 kg). Male leopards take no part in caring for cubs. The **leopardess** is their sole provider and caregiver.

A leopard cub opens its blue eyes six to 10 days after birth. Every few days, a mother leopard moves her cubs to a new den. This keeps them safe from predators. For the first few months, cubs survive only on their mother's milk. They wait patiently when she leaves them to hunt. Under their mother's watchful eye, cubs wander short distances from the den. They begin to sample small bites of meat their mother brings back from hunts. In this way, cubs develop a taste for the prey they'll soon learn to hunt on their own. A mother leopard and her cubs stay together for two years.

6 WEEKS

- Coat thickens and is now more golden in color
- Black spots are more visible
- Teeth have erupted
- Practices stalking and pouncing with sibling or mother

CLOSE-UP

Lazy days

Leopards are **nocturnal** animals that hunt at night. Large eyes help them see well in the dark. During the day, leopards enjoy resting on a tree limb. They like to dangle their legs off the sides.

FEATURED FAMILY

First Meal

Totally dependent on his mother, the newborn male Amur leopard nurses soon after birth. He nuzzles in, finds a nipple, and drinks his mother's rich milk. In the early days after birth, his mother will spend a lot of time snuggling, cleaning, and nursing him. After about a week, the cub's eyes open. He's more alert now, and he's steadier on his feet. For the next few months, his mother's milk will provide his only nourishment. Then, he'll begin to nibble on hare or sika deer scraps his mother brings home from her hunts.

Leopards ENJOY spending the day RESTING on a TREE LIMB.

8 WEEKS

- Eats scraps of meat from mother's hunts
- Coat covered in rosettes
- Weight has doubled since birth

12 WEEKS

- Weaned
- Follows mother when she hunts
- Stalking, pouncing, and climbing skills improve with practice

CLOSE-UP

Skilled climbers

Leopards are excellent climbers. They are strong enough to carry carcasses three times their own weight up into trees. They stash prey on branches to keep it away from other predators. This practice, called caching, allows a few days of undisturbed feeding.

CHAPTER TWO

EARLY ADVENTURES

For weeks, leopard cubs practice stalking and pouncing behaviors as they play. Once they are **weaned**, they follow their mother into the forests and grasslands to hunt for real. Their eyes have turned from blue to yellow. Their senses are keen. Excellent sight and hearing help them sneak up on prey unnoticed. As cubs grow, their coats get more golden in color. Their black rosettes become more visible. This color pattern serves as camouflage to hide leopards from both predators and prey.

Leopard cubs watch as their mother hunts. Crouching nearby, they mimic her actions to practice their hunting skills. A leopardess barely makes a sound as she stalks. She pounces on prey animals and strangles them with a crushing bite to the neck.

4 MONTHS

- Stalks and kills small prey for food
- Sticks close to mother and sibling

6 MONTHS

- Stalks and kills medium-sized prey
- Now less likely to become prey for larger animals

CLOSE-UP

Black panthers?

Some leopards overproduce a pigment called melanin. Melanin makes their fur black. These cats are often called black panthers, but no such species actually exists. The black fur masks the leopard's spots, but the spots are still there.

FEATURED FAMILY

Look Who's Stalking

As he grows, the Amur leopard cub watches his mother closely. He sees how she crouches so low that her belly fur scrapes grass. He notices how she moves each paw through the weeds without a sound. One day, he witnesses a surprise attack on a roe deer as his mother pounces and sinks her teeth into the animal's neck for the kill. Mimicking her, the cub and his sister stalk and pounce as they play. The female cub crouches low in the grass and sneaks up on her brother. She pounces on his back, and the twins tumble in the grass. They are practicing skills they will soon need for hunting.

Using her strong jaws and massive strength, a mother leopard will carry prey to a nearby tree. She climbs high and wedges the carcass between two tree limbs. The mother eats first. Then, her cubs make their way to the branch and take their turn. Leopards often eat high in trees to prevent scavengers from taking their food. This practice also protects cubs from predators that smell the fresh kill.

By imitating their mother, leopard cubs soon learn to hunt on their own. They begin by catching insects and rodents. They advance to larger prey. With each kill, young leopards develop the ability to be completely silent as they stalk animals through grasses and underbrush. Their pounces become more powerful and accurate. By the time they are 18 months old, they are skilled hunters.

12 MONTHS

- Hunts larger prey with sibling's help
- Hauls large prey high into a tree before eating

18 MONTHS

- Capable of taking down large prey
- Hunts independently, but remains with mother

CLOSE-UP

Watch the tail

The tip of a female leopard's tail has a white underside. It is easy to see, even in low light. Scientists think that a leopardess curls her tail up when she walks to communicate with her cubs and say, "Follow me."

FEATURED FAMILY

Give It a Try

Now four months old, the Amur leopard cub stalks a squirrel. His mother stays nearby. He steps on a twig, and it snaps loudly. Alerted by the noise, the squirrel scampers up a tree and escapes. The cub is not discouraged. He spies a mouse and crouches to stalk his prey. Just as he's ready to pounce, his mother yelps a warning cry. Four wolves rush toward the leopards. Following his mother's lead, the cub dashes to the safety of a tree and leaps onto its lower branches. He is still hungry, but he's safe. He will try again another day.

Leopard cubs practice STALKING and POUNCING behaviors as they play.

24 MONTHS

- Leaves mother to establish own territory
- Males and females hunt and live independently
- Leads a solitary life
- Males weigh 99 pounds (45 kg)
- Females weigh 88 pounds (40 kg)

CLOSE-UP

True leopards

Although *leopard* is part of their names, snow leopards, clouded leopards, and leopard cats are not true leopards. The sizes and structures of their skulls and throats are different from leopards. Scientists have determined that these animals are not closely related to leopards.

CHAPTER THREE

LIFE LESSONS

When they are about two years old, leopard cubs leave their mothers to lead a solitary life. Each adult leopard establishes its own territory. Often, females share part of their mother's territory. The territories of males and females may also overlap. The size of a leopard's territory can range from four to 40 square miles (10 to 100 sq. km).

Leopards make sounds to communicate and to show their emotions. Growls, hisses, and snarls express aggression. Males roar to defend their territory. Females roar to attract the attention of males. A female's soft grunts call her cubs. Cubs mew in return. Both males and females make a noise that sounds like sawing wood. Scientists call the sound a "saw." A saw announces the boundaries of a leopard's territory. It is also used for communication between males and females. Another sound, called a "chuff," is made during friendly encounters. To make it, a leopard pushes air through the nose and lips.

(26) MONTHS

- Reaches sexual maturity
- Increases scent marking, calling, and roaring
- Most active at night, but can be seen in daytime
- Rests and sleeps high in trees

Leopards also communicate through scents. They mark their territory by leaving their urine or feces in the area. Other leopards smell it and stay away. Leopards also send messages by rubbing their faces against trees or other objects. This releases an odor from scent glands in the cheeks. When leopards make claw marks on tree trunks, they deposit a scent released from glands in their paws. Other leopards pick up the scents and know who has been there.

Leopards reach sexual maturity between the ages of two and three. A female leopard often begins courtship. She seeks out a male by traveling to his territory. She moves back and forth in front of the male, rubs her head against him, and twitches her tail. After the female becomes pregnant, the **gestation** time for the cubs is about three months.

FEATURED FAMILY

This Is How It's Done

The leopard mother crouches low to stalk a pair of hares. The cub follows, watching his mother's white-tipped tail to help him navigate through the thick brush. The leopardess falls back and lets the cub move ahead. Stalking slowly and as silently as he can, he creeps closer. He waits a moment before pouncing on the nearest hare. He **suffocates** the animal with a single bite to the neck. The cub is almost done with his meal when he catches a familiar smell. A Siberian tiger is nearby! Leaving his food behind, the cub and his mother run to the safety of their den where the tiger can't follow.

CLOSE-UP

Spots

To tell the difference between cheetahs, jaguars, and leopards, look closely at their spots. A cheetah's spots are solid black. Jaguar splotches are larger than leopard spots, and they have one or more black dots in the center of each splotch. Leopard rosettes are smaller with no black dots in the middle.

3 YEARS

- Females give birth to first litter of cubs
- Females care for cubs
- Males take no part in cub care

4 YEARS

- Males hold established territories
- Females spend half their adult lives caring for cubs

Most leopards live for 12 to 17 years. Mastering the art of hunting equips them for survival. However, even the most skilled hunters must remain alert for dangerous predators that may shorten their lives.

CLOSE-UP

Movement

Leopards can run 36 miles (58 kilometers) per hour. They can jump forward 20 feet (6 m) to pounce on prey. They can leap 10 feet (3 m) straight up to climb trees and boulders and to catch birds out of the air.

FEATURED FAMILY

Practice Makes Perfect

The young leopard has grown strong and skillful. Weeks after leaving his mother, he explores a section of the forest, hoping to find a territory to call his own. He is focused on scent markings left on the trees, but he does not miss a slight stirring in the leaves. In an instant, he sees and smells a sika deer and crouches low. He lifts and places his paws carefully into the snow to keep it from crunching. Then, at just the right time, he pounces across a stream and wraps his powerful front legs around the animal. One deep bite to the neck completes the kill. The leopard easily drags his fallen prey to a nearby tree and hauls it up to drape across a limb. He eats his fill.

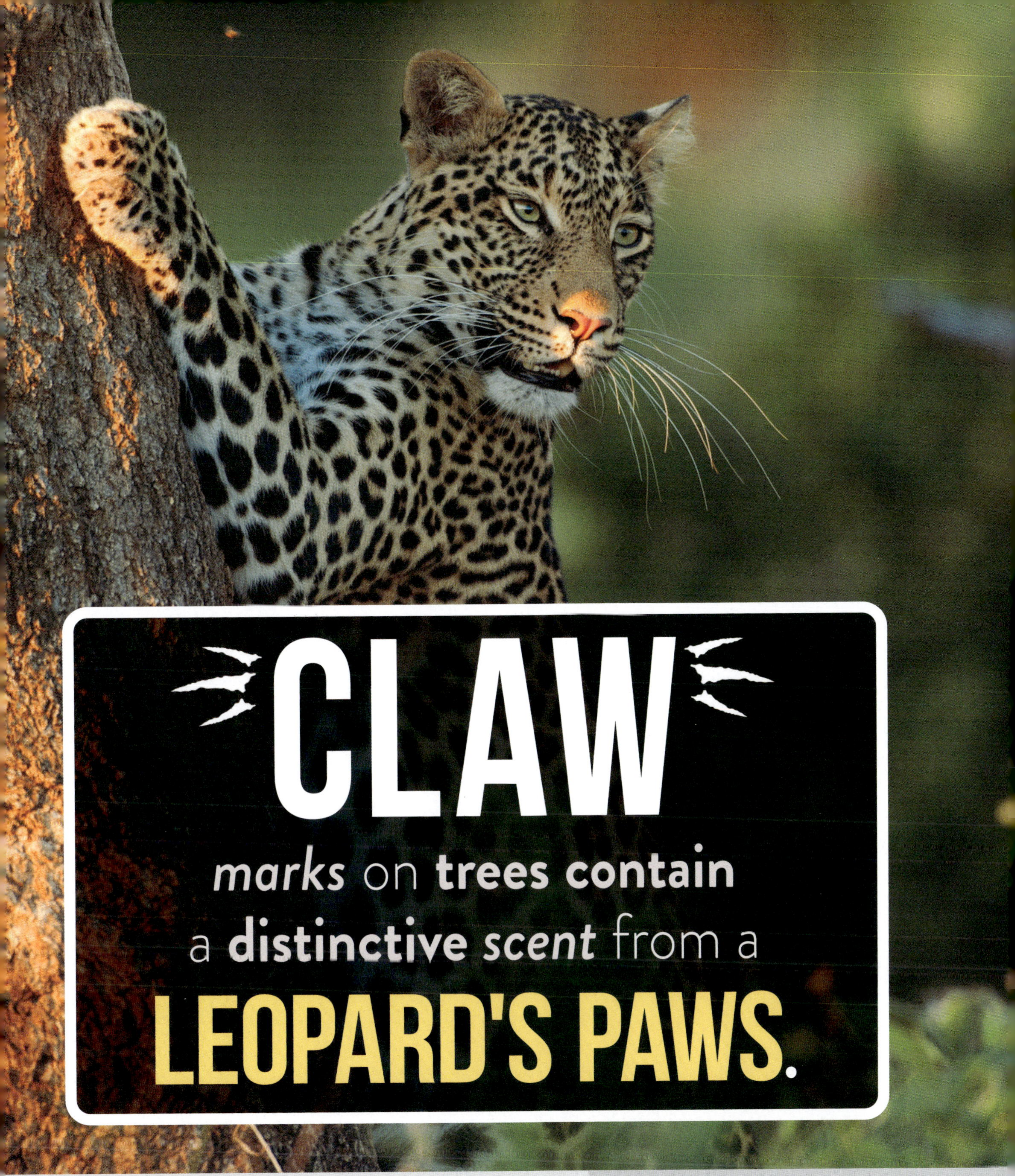

CLAW *marks* on **trees contain** a **distinctive** *scent* from a **LEOPARD'S PAWS.**

(12) YEARS

- Females give birth to final litter
- Cub survival declines with age of mother

(16) YEARS

- End of life

CHAPTER FOUR

LEOPARDS LOSING GROUND

Leopards are struggling to survive. Around the world, loss of habitat has decreased their numbers. All leopard subspecies are classified as threatened or endangered. Russia's Amur leopard is the rarest and most vulnerable subspecies. **Conservationists** estimate that less than 100 live in the wild.

Hundreds of years ago, leopards roamed a much larger portion of Africa, Europe, and Asia. However, people have claimed much of their habitat and used it for farms and other types of development. This reduces the land available for leopards and shrinks their territories. Less wild habitat also means fewer prey animals for leopards.

Forcing leopards into smaller territories often brings them closer to humans. Leopards may begin to prey on farm animals. Sheep or cattle farmers sometimes kill leopards to protect their livestock. Illegal hunting for leopard skins, teeth, and meat also causes the deaths of many leopards.

Many groups are trying to protect leopards. The African Wildlife Foundation works closely with farmers to protect their animals without harming leopards. In some places, this organization builds fences that protect livestock from leopards. The Greater Kruger Leopard Conservation Science Project places GPS (global positioning system) collars on leopards. This helps researchers study their behaviors and movements.

The WildCats Conservation Alliance and Wildlife Conservation Society work together in China. Researchers set up camera traps to record the number of Amur leopards. They mount cameras on trees where leopards roam. The cameras snap pictures when movement sensors and heat sensors detect the presence of wildlife. Wildlife biologists can look at the pictures and identify leopards by their spots. Each leopard's rosettes are unique, much like human fingerprints. Information from the cameras helps conservationists create safe places for the big cats to live.

Many zoos have leopard breeding programs, including California's San Diego Zoo and England's Dartmoor Zoo. These programs try to increase leopard populations. At some zoos, new Amur leopard cubs have been born recently. The Highland Wildlife Park in Scotland built an Amur leopard habitat designed to encourage leopard breeding with very little human contact. This enables the park to introduce cubs into the wild when they mature.

Although many efforts are underway to save leopards around the world, there is much more to do. Learning about leopards can help people make changes to protect these beautiful, magnificent cats.

FAMILY ALBUM

SNAPSHOTS

The **Sri Lankan leopard** has no competition from other big cats. These leopards are the apex predators, or top predators, on the island of Sri Lanka.

Found only on the densely populated island of Java in Indonesia, the **Javan leopard** is critically endangered. Conservation efforts are underway to protect this last remaining large carnivore on Java.

Indochinese leopards are often melanistic, or black, in some parts of their range in southwest Asia.

In 2017, the Cat Classification Task Force of the Cat Specialist Group ruled that the **North China leopard** subspecies should be classified together with the Amur leopard subspecies. Some wildlife biologists disagree with the change.

The **Persian leopard** is the largest of all the leopard subspecies. Their numbers are threatened by illegal hunting and loss of habitat.

The **Arabian leopard** is critically endangered. It has one of the smallest populations of all the subspecies, with fewer than 200 animals left in the wild.

Indian leopards have been known to hunt people. Their history as a dangerous predator prevents some people from supporting leopard conservation in India.

African leopards in sub-Saharan Africa often take over aardvark burrows to use for their dens.

In an effort to save **Amur leopards**, the country of Russia has protected almost 650,000 acres (263,000 ha) of the leopard's habitat and called it Land of the Leopard National Park.

WORDS to Know

conservationists people who try to save and protect Earth's natural resources, including wildlife

gestation the period of time it takes for a mammal's offspring to develop inside the mother; pregnancy

leopardess a female leopard

mimicking imitating or copying

nocturnal active at night and at rest during the day

opportunistic taking advantage of whatever opportunities or resources are available

stalk to pursue secretly and silently

subspecies types or varieties of living things within a species

suffocates kills an animal by stopping its breathing; smothers or strangles to death

weaned no longer getting food by nursing from a mother animal; at a stage of development in which an animal must feed itself

LEARN MORE

Books

Eszterhas, Suzi. *A Leopard Diary: My Journey into the Hidden World of a Mother and Her Cubs*. Toronto: Owlkids, 2022.

Grady, Tyler. *Leopard: Fascinating Animal Facts for Kids (This Incredible Planet)*. Sarasota, Fla.: Dylanna Publishing, 2022.

Winter, Steve. *The Ultimate Book of Big Cats: Your Guide to the Secret Lives of These Fierce, Fabulous Felines*. Washington, D.C.: National Geographic, 2022.

Websites

"14 Fascinating Facts About a Leopard's Life Cycle." African Wildlife Report. https://africanwildlifereport.com/14-fascinating-facts-about-a-leopards-life-cycle

"Leopard." National Geographic. https://www.nationalgeographic.com/animals/mammals/facts/leopard

"Leopard (Panthera pardus) Fact Sheet: Summary." San Diego Zoo Wildlife Alliance Library. https://ielc.libguides.com/sdzg/factsheets/leopard/

Documentaries

Joubert, Dereck and Beverly. *Eye of the Leopard*. National Geographic, 2006.

Leonard, Steve. *Animal Kingdom: Cheetahs and Leopards*. Real Wild, 2011.

Path of the Hunter. Wild Planet, 2020.

Note: Every effort has been made to ensure that any websites listed above were active at the time of publication. However, because of the nature of the Internet, it is impossible to guarantee that these sites will remain active indefinitely or that their contents will not be altered.

Visit

KRUGER NATIONAL PARK

Established as a wildlife refuge in 1898, Kruger offers self-drive safaris, guided game drives, and overnight stays.

Skukuza 1350
Mpumalanga, South Africa

MAASAI MARA NATIONAL RESERVE

Take a hot-air balloon adventure over the wildlife reserve or a 4 x 4 safari adventure to see it at ground level.

C13 Lolgorien
Narok, Kenya

ROOSEVELT PARK ZOO

Be sure to look for enrichment toys and items offered to the zoo's leopards and other animal inhabitants.

1219 Burdick Expressway East
Minot, ND 58701

SAN DIEGO ZOO

Look for the twin Amur leopard cubs on exhibit.

2920 Zoo Drive
San Diego, CA 92101

INDEX